EXPLORING CAREERS WITHOUT COLLEGE

VOCATIONAL CAREERS IN TECHNOLOGY

by Cynthia Kennedy Henzel

BrightPoint Press

San Diego, CA

an imprint of ReferencePoint Press, Inc.
Printed in the United States

For more information, contact:
BrightPoint Press
PO Box 27779
San Diego, CA 92198
www.BrightPointPress.com

LIBRARY OF CONGRESS CATALOGING-IN-PUBLICATION DATA

Name: Henzel, Cynthia Kennedy, author.
Title: Vocational careers in technology / by Cynthia Kennedy Henzel.
Description: San Diego, CA: ReferencePoint Press, 2026 | Series: Exploring careers without college | Includes bibliographical references and index. | Audience: Grades 7–9
Identifiers: ISBN 9781678212766 (hardcover) | ISBN 9781678212773 (eBook)
The complete Library of Congress record is available at www.loc.gov.

CONTENTS

THE TECHNOLOGY INDUSTRY AT A GLANCE

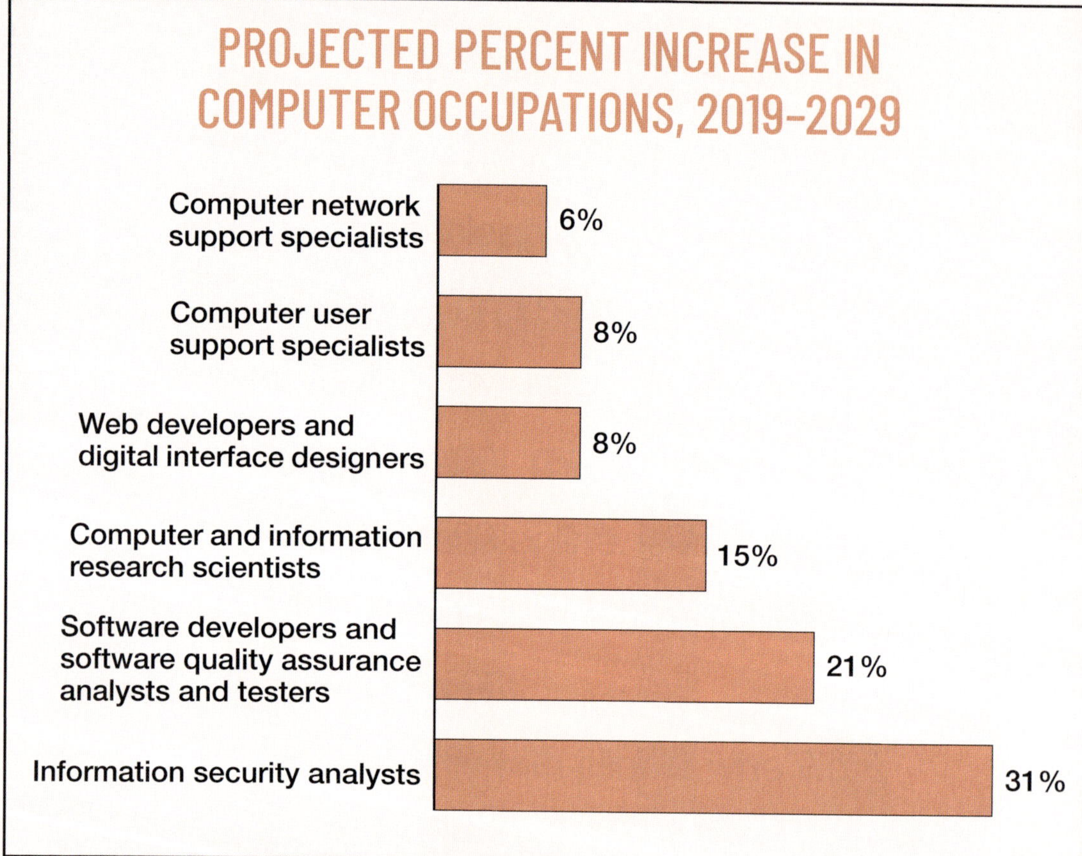

Source: Alan Zilberman and Lindsey Ice, "Why Computer Occupations Are Behind Strong STEM Employment Growth in the 2019–29 Decade," US Bureau of Labor Statistics, *January 2021, www.bls.gov.*

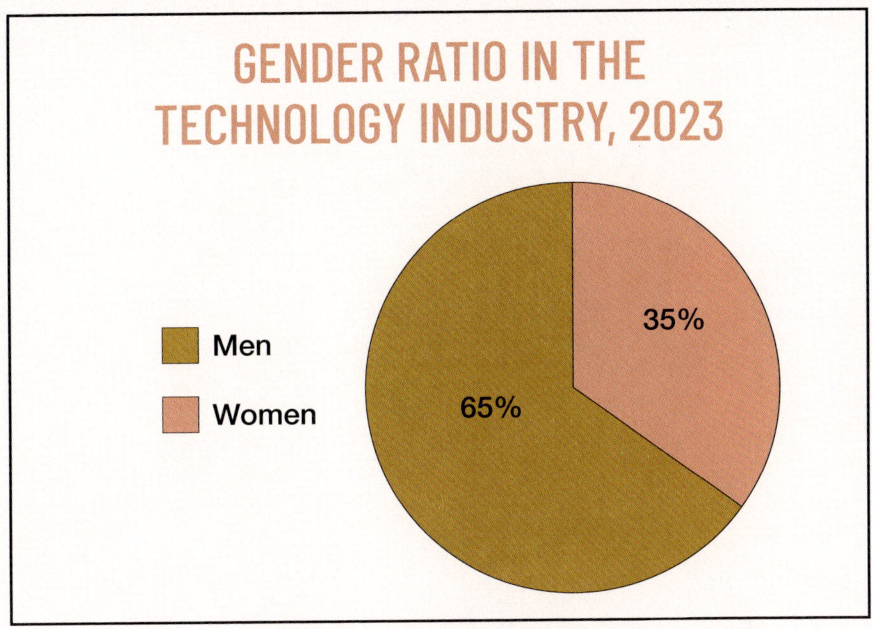

Source: "Women in Tech Stats 2025," WomenTech Network*, 2025, www.womentech.net.*

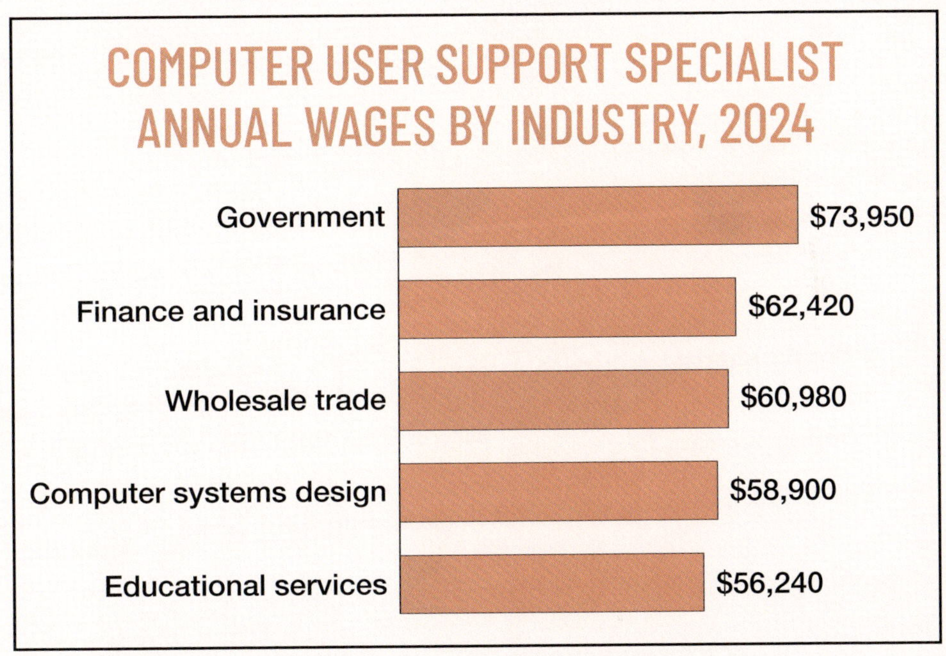

Source: "Computer Support Specialists," US Bureau of Labor Statistics*, August 28, 2025, www.bls.gov.*

WHAT IS THE TECHNOLOGY INDUSTRY?

New Berlin Eisenhower is a high school in Wisconsin. The heads of the school wanted to teach students about trades. They bought a robotic arm. The arm was used in a class about how machines work. This is how Aren Schiek became interested in robotics.

In the class, Aren used the robotic arm. This type of robot is often used in factories. First he used the robot to

Many US schools have a robotics program in which students learn how to design, build, and program robots.

During robotics competitions, teams have their robots do tasks such as going around obstacles or moving objects.

draw circles. Then he learned more advanced programming. Aren worked with two classmates. They entered robotics competitions. In 2021, they won the Wisconsin Manufacturing and Technology Show. The next year, they won another competition. It was called Wisconsin SkillsUSA.

Aren's success led to another opportunity. Viking Masek was a local robotics company. It needed a new production intern. Aren was still in high school. But the company was impressed with his skills. So he got the job. Soon, Aren was doing hands-on work. He did electrical wiring. He also put together some of the systems the company built. This allowed Aren to keep learning.

WORKING IN THE TECHNOLOGY INDUSTRY

The technology industry involves working with computer systems. It also involves services that use computers. Some workers build computers. They build computer parts, too. Others develop software.

Some experts estimate that there are about 585,000 software and information technology service companies in the United States.

Information technology (IT) is a key part of the industry. It includes people who create websites. Many people in IT store and study data. Others help protect it. Some run computer **networks**. There are also IT workers who design computer games.

The technology industry changes quickly. Areas such as robotics and artificial intelligence (AI) see rapid improvements. Technology specialists might work in different fields. These include health care or renewable energy. Another is virtual reality. The tech industry offers many opportunities for people who like working with computers.

CHAPTER ONE

CNC TOOL PROGRAMMER

Computer numerically controlled (CNC) tool programmers write computer code. This code helps control **automatic** machine tools. These tools cut and shape metal or plastic parts. The parts are used in many industries. These include transportation and health care.

CNC programmers write code in many computer languages. The most common is G-code. G-code tells a machine how

CNC programmers input code through a control panel, which is considered the brain of a CNC machine.

to move. It also tells it how fast to move. Programmers use computer-aided manufacturing (CAM) software. This can create G-code. Understanding G-code helps programmers solve problems.

CNC Tool Programmer

Minimum Education: High school diploma
Personal Qualities: Computer skills, strong math skills
Certification and Licensing: CNC certification is optional, but some states require a license.
Working Conditions: CNC tool programmers spend most of their time on computers. They sometimes travel to check machines to ensure they are working properly.
Average Salary: $63,440 (2023)
Number of Jobs: 28,300 (2023)
Future Job Outlook: 15 percent growth expected between 2023 and 2033

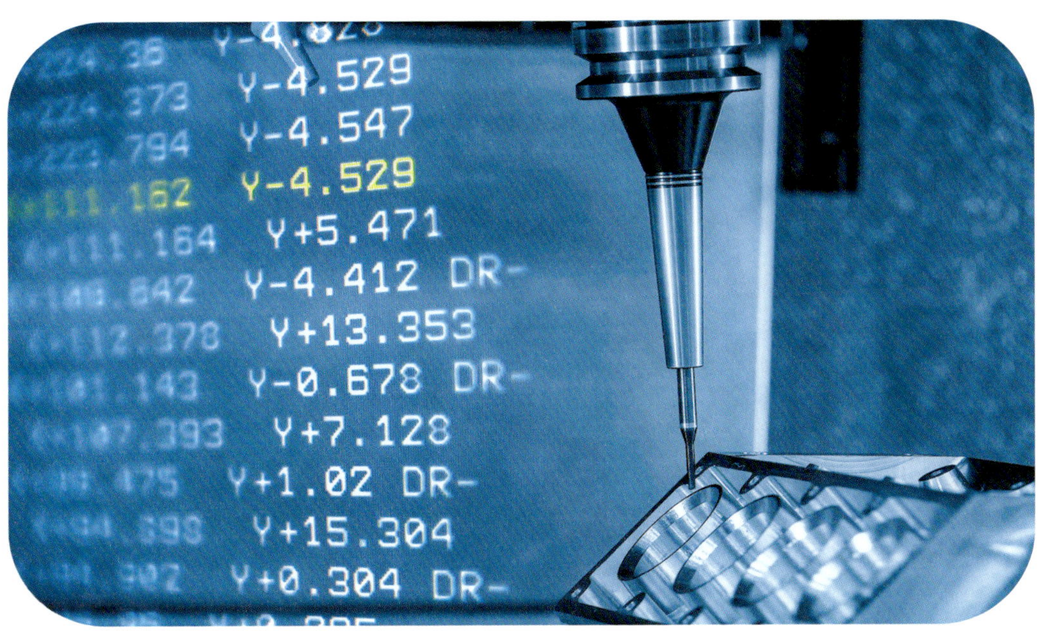

G-code is short for geometric code. It uses letters and numbers to tell machines information.

Automatic machine tools often use machine learning (ML). This means they are programmed to recognize patterns and take action. ML tools can improve. To do this, they need more data. They also need more experience. For example, ML machines can predict when a machine needs to be fixed. This keeps machines running efficiently.

CNC programmers talk to workers who use machines. This helps them learn how well the program works. They monitor the equipment as well. This helps them make better programs. Some programmers also set up equipment. They may operate it and fix it, too.

CNC programmers must be organized. They need to put program instructions in

Hands-on training is an important way for a person to learn the skills needed to become a CNC programmer.

the correct order. Programmers need good observation skills. This helps them find problems. Math is important, too. It can help solve some problems.

GETTING STARTED

CNC programmers need a high school diploma. Math and physics classes are helpful. These subjects are used when writing code for machines.

Many students go to a vocational school. Certification programs teach CNC programming basics. Students learn to write code for CNC machines. Vocational programs take 6 weeks to 1 year. Different schools offer CNC programming for certain industries. Students should choose an industry that interests them.

Some companies train on the job. Apprenticeships are available, too. These programs allow students to learn from a professional. Students learn in the classroom, too. Apprenticeships may last 2 to 4 years. Apprentices work with experienced CNC programmers. They often learn skills needed to work at the company that employs them.

Experience helps people move up in the field. Experienced CNC programmers can lead teams of other programmers. They might develop **specifications** for projects. Or they might **troubleshoot** issues. Some even help train new staff. Programmers must always learn new technology, such as AI. Todd English is the vice president at a CNC manufacturing company. He says,

"If you don't keep up with all of the new technologies, you won't maximize your production. Everyone is looking to become more efficient at what we do."[1]

ON THE JOB

CNC programmers work on different projects. So they must learn what pieces they need the machine to make. They look

Artificial Intelligence and Machine Learning

Artificial intelligence and machine learning are not the same. ML is part of AI. AI is the ability of computers to do tasks usually done by humans. Some AI can recognize human language. Other forms of AI are modeled after the human brain. These are known as neural networks.

CNC programmers must know how to read blueprints, which are detailed plans that show what a part should look like.

at drawings and building plans. They read the specifications. Programmers figure out how big the parts should be. Then they think about the commands needed. Commands are instructions. They tell the machine what to do. This might be cutting away material to create a shape. Or it might be drilling holes. Programmers must also decide the order of the commands.

Programmers then select the machine's cutting tools. These are needed to make the part. Programmers also decide how fast the machine will work. This information is used to write a program. This tells the machine what to do.

After the program is written, it is tested. Programmers set up the machine. They install the program. They watch it work. And they make sure the finished product meets specifications. Sometimes there are problems. Programmers make changes to the code. Then they test again. This continues until all problems are solved.

Finally, the programs are stored electronically. Programmers write instructions on how to set up the machine. They also explain how to use the program. These instructions are used by the workers who run the machine.

CNC programmers talk with other workers. Workers may know how to make the job easier or more efficient. These ideas might become part of the program. Many programmers also maintain equipment. They troubleshoot problems as well.

Ricky Strader is a CNC programmer. He says the work is challenging. Sometimes, making a part takes a lot of time. The technology is always changing, too. But Strader says, "In the end, it's rewarding to be able to step back and see the end product."[2]

FIND OUT MORE

National Institute for Metalworking Skills
www.nims-skills.org
The National Institute for Metalworking Skills (NIMS) website has information on training and apprenticeships. The NIMS site also offers information on certifications.

National Tooling and Machining Association
https://ntma.org/training-and-education
The National Tooling and Machining Association (NTMA) provides members with networking opportunities in the CNC trade. The NTMA website includes information on apprenticeships and on-the-job training.

CHAPTER TWO

COMPUTER SUPPORT SPECIALIST

Computer support specialists (CSSs) help people with computer problems. CSSs often work with a customer in person. They might also work through email or over the phone. Some even work in an online chat. Specialists try to solve problems. They might fix the issue themselves. Or they explain to customers how to fix it.

CSSs also repair computer equipment. They may install and update software.

CSSs answer customer questions and troubleshoot computer problems such as internet or printer connection issues.

They maintain equipment and networks. A CSS might set up new equipment. They show people how to use an updated network. Their work helps organizations run smoothly.

Computer Support Specialist

Minimum Education: High school diploma
Personal Qualities: Good customer-service, problem-solving, and computer skills
Certification and Licensing: Certifications in different technologies recommended
Working Conditions: Computer support specialists work full-time with computers and customers. They may need to be available nights and weekends for computer and network support.
Average Salary: $61,550 (2024)
Number of Jobs: 892,000 (2023)
Future Job Outlook: 6 percent growth expected between 2023 and 2033

CSSs work in many industries, including finance, health care, and education.

Specialists must have good technical skills. They need to know how to solve computer problems. Good communication skills are key. So are customer-service skills. CSSs must listen to the customer. Patience helps when solving issues.

GETTING STARTED

CSSs need a high school diploma. Math and computer science classes are useful. Most students take courses at

Some CSSs work in an office setting, while others work from home.

a trade school. These courses include networking, server administration, or information security. Many of these classes lead to certifications.

Some certifications provide a basic introduction to the field. One is the Google IT Support Professional certificate.

Companies such as Microsoft or Oracle offer training and certifications on their software.

Experience helps CSSs advance their careers. Continued education does as well. CSSs must know about new security threats. Hardware and software technology advances quickly. Successful CSSs learn about these advancements. They also earn advanced certifications in different areas.

ON THE JOB

There are several types of CSSs. One is a help desk technician. They work through email, online chat, or the phone. They listen to the user's problem. They may ask questions. Techs talk users through how to solve the problem. Some help desk techs work for large organizations to support customers. Others work for technology companies. They help users with specific software issues.

Ben Gof is a help desk technician. He explains that good CSSs know how to figure out problems. Gof says:

Sometimes, you will need to dig in and find out why the normal solution

Resetting Passwords

CSSs solve issues with account passwords. Users may type passwords in wrong. Or users might forget them. Passwords also expire. CSSs can reset passwords. They encourage users to choose passwords that are easy to remember. Passwords should also be hard for others to guess. This keeps computers secure from criminals.

Support specialists might talk to others on their team to figure out solutions for difficult problems.

isn't working. But that is also a part of the satisfaction and enjoyment, which happens when you solve that really tough problem that no one else could figure out![3]

Another CSS role is a technical support specialist. This job involves supporting customers within an organization. Some work in schools. Others might work at health care facilities. These places have software designed for their particular industry. Specialists must be familiar with that software.

Technical support specialists do many tasks. They may begin the day by checking

Server rooms are places that hold a company's network equipment and computer servers.

for computer **viruses**. They respond to emails from workers having computer problems. Specialists might study a new software program, too. They need to know how it works. And they need to know how to install it. Finally, specialists might prepare for a software update. This might happen overnight when others are not working.

David Thibodeau is a support specialist. He enjoys the variety in his job. He says:

> *I train people on software and systems. I take calls from people who are having difficulty [with] software/hardware issues. I help people track down lost files and rebuild computers to get them up and running again. Every day is a bit different. The job is really cool that way.*[4]

Network support specialist is another type of CSS. Network specialists do not work directly with customers. They troubleshoot and maintain local area networks. These include all the devices within a single area, such as a business. Network support specialists also work with wide area networks. These cover a large area, such as a city. Network specialists work on internet networks, too.

The job of network specialists includes backing up data. They also provide network access to workers. Specialists check networks to make sure they are running smoothly. They ensure new equipment works with the existing network. They also keep the network secure.

FIND OUT MORE

Computing Technology Industry Association
www.comptia.org/en-us/explore-careers/job-roles/tech-support-specialist
The Computing Technology Industry Association (CompTIA) offers information on roles in the computer industry. CompTIA provides training and certifications at all levels. Its website features blogs about jobs in technology.

Microsoft
https://learn.microsoft.com/en-us/training
Microsoft offers free online training for all types of computing. The site has a Student Hub section that has information on technology competitions as well as training for students in many types of software.

WEBSITE DEVELOPER

Website developers create websites for people and businesses. Websites have many uses. These include e-commerce. This is selling products or services online. Some sites provide information. Gaming sites provide entertainment.

Developers meet with clients who want a website. They talk about why the site is needed. They discuss what should be

Website developers make sure websites can work on different devices, such as computers, tablets, and smartphones.

on it. And they discuss how the site will work. Developers often work with website designers. Designers and developers do different tasks. Web designers focus on how a site looks. They choose colors or logos. They edit graphics and write content.

Website Developer

Minimum Education: High school diploma

Personal Qualities: Strong computer skills, good communication skills, attention to detail

Certification and Licensing: None required

Working Conditions: Website developers work mainly on computers, and they interact with clients. Some work in an office, while others work from home.

Average Salary: $95,380 (2024)

Number of Jobs: 222,600 (2023)

Future Job Outlook: 8 percent growth expected between 2023 and 2033

Talking to clients can help web developers work toward achieving the goal or purpose of a website.

Web developers write the code that makes the website work.

There are two main types of developers. Back-end developers create the main structure of a website. This is where data is stored. It is also where business sales are processed. Back-end developers ensure the website is secure. They also make sure it functions well.

Front-end developers create the parts of sites that users see. These include menu bars and links. This makes content easy

Web developers often work with other people, such as web designers, to plan out the structure and features of a site before they begin coding.

to find. Front-end developers add text and graphics made by designers or clients. They make a site look good as well.

GETTING STARTED

Website developers need a high school diploma. Coding and computer classes are useful. Many students learn the basics of web development online. These classes can take about 2 months to complete.

Websites Without Coding

Some online services can build sites without coding. One example is Wix. These services are often used by people building their own websites. However, they can be limiting. People might have only a few layouts to choose from. The site's functions may be limited, too.

Some students attend a vocational school. They practice creating websites. These can be used in a developer's **portfolio**.

Web developers need to learn computer coding. Computer coding is writing instructions to tell a computer what to do. These instructions make a website function. Code is written in coding languages. Back-end developers might use Structured Query Language, or SQL. It is used to store and process information.

Hypertext Markup Language (HTML) is a common language. It is used in front-end development. HTML provides the main structure of a web page. Cascading Style Sheets (CSS) is another language. CSS is used to change design elements. Some developers use JavaScript, too.

More than 8,000 coding languages exist today. Some commonly used languages include Python, Java, and C++.

ON THE JOB

Web developers at small companies may have full responsibility for a website. They register domain names. A domain name is a company's web address. Developers work with website hosting companies to put a site on the internet. They design, develop, and maintain the site.

Being detail oriented and having good problem-solving skills are important qualities for web developers.

Developers who work for large companies often work with a team. They may specialize in one part of the development process. Sam works on a four-person development team. He spends most days working on new and ongoing projects. “If I have time, I like to end the day by watching a video tutorial,” Sam says.[5] This helps him learn about new developments in the field.

Developers must create websites that are easy for people to use. Pages must load quickly. Slow links can make users leave a site. Websites must work on different devices. Developers load new websites onto a testing site. Team members review the code. They search for problems. They make sure links work. Once a site is running, developers continue to maintain it. They update it as needed.

Freelance website developers run their own businesses. They make their own schedules. They choose the projects they want to work on. Most freelancers work from home. They buy their own equipment and supplies. Freelancers often design and develop websites. They also find their own clients. Most maintain their own websites.

The salary range for website developers varies. The US Bureau of Labor Statistics says some developers make less than $48,560. The top salaries can be more than $162,870. Salaries vary by location and industry. A developer's education and skills are an important factor, too.

Brandon Arnold is a web developer. He believes continuing to learn is an important part of this job. Arnold says, "The key to being a successful web developer is to stay up-to-date on modern trends and technologies."[6]

FIND OUT MORE

Coursera

www.coursera.org/articles/web-developer

The Coursera website highlights different careers, including web developer. This article includes information on the skills needed to become a web developer and the tasks people do in this career field.

Udemy

www.udemy.com/courses /search/?src=ukw&q=website+design

Udemy hosts educational videos from professionals in many industries. It has a large collection of courses offered by website development and design professionals.

CYBERSECURITY ANALYST

Cybersecurity analysts ensure computer data and networks are secure. They usually work for organizations. Analysts protect a company's private information. They make sure computer systems are not harmed by viruses or illegal software.

Cybersecurity analysts install firewalls. Firewalls are security barriers. They keep an organization's data from being stolen. **Hackers** often try to steal important data.

Cybersecurity analysts use a variety of digital security tools, such as firewalls and antivirus software, to help protect data.

Analysts install encryption software. This ensures data can be read only by people with proper access.

Analysts develop security rules for workers. This prevents workers from accidentally causing a security **breach**. If a

Cybersecurity Analyst

Minimum Education: High school diploma, some college preferred but not essential

Personal Qualities: Problem-solving and analytical skills, communication skills

Certification and Licensing: Certification preferred, such as Security+

Working Conditions: Cybersecurity analysts mainly work on computers. They might work overtime or be on call for security issues.

Average Salary: $124,910 (2024)

Number of Jobs: 180,700 (2023)

Future Job Outlook: 33 percent growth expected between 2023 and 2033

breach does occur, analysts investigate. Then they fix the problem.

Cybersecurity analysts must have programming skills. They should be detail oriented. Analysts must know about current industry trends. They also need to think about future industry changes. Analysts try to prevent new types of attacks before they happen.

GETTING STARTED

Security analysts need a high school diploma. They should take classes in math and computer science. Most employers prefer students to have a college degree. However, people can work in this field without a degree.

Some people do cybersecurity analyst internships. These are opportunities in which students can gain hands-on experiences within a career field.

Students without a degree must work hard to learn about security. They should also gain experience in the field. Security professional Vandana Verma says, "Keeping yourself up to date is the key. Someone who has curiosity about every aspect of technology is probably the best suited person."[7] Talking to industry professionals can be helpful. This is also one way to learn about job opportunities.

Some analysts begin by taking online classes. They learn cybersecurity basics. One course is Security Analyst Fundamentals. It is offered by the company International Business Machines (IBM). This course takes about 1 month to complete. Students learn how to prevent security threats and what to do during attacks.

Employers prefer analysts to be certified. One certification is Security+. It shows that an analyst knows how to secure networks and devices. People with more experience can earn higher-level certifications.

ON THE JOB

Analysts must prevent network attacks. This includes ransomware attacks. These are when hackers access a network.

They block legal users from getting into the system. Then they demand payment to give access back. Another threat is social engineering attacks. These trick users into doing things online that let a hacker into a system. An example is clicking a fake link. Analysts also watch for threats from inside an organization. Employees can steal private data, too.

Cybersecurity analysts look for and prevent problems. They use software that scans systems for threats. They keep security software up to date. Analysts make sure other workers are following security rules. They also report possible problems to company leaders.

Educating employees about security is an important part of the job. Analysts develop training programs. These teach workers to avoid becoming a security risk. Analysts make sure people follow these rules.

Even with security, breaches can still happen. Analysts develop emergency and recovery plans. For example, they might shut down email systems during an emergency. This keeps viruses from spreading to other computers. By working

White Hats

Penetration testers are a type of cybersecurity analyst. They are sometimes called white hats. This comes from a tradition in old Western movies. The good guys usually wore white hats, while the bad guys wore black. White hats act like hackers. They are hired by companies to try to hack into computer systems. This helps companies figure out how to improve their security.

quickly, analysts can protect a system from further damage. Then they identify and stop the attack. Finally, analysts check that all data is safely backed up.

Being a cybersecurity analyst can be stressful. Hackers can attack at any time. Analysts may be on call during nights or weekends. Keeping up with new technology can also be difficult. But cybersecurity analyst Lance Spitzner thinks this is the best part of the job. He says:

> *What I love about cybersecurity is how it's a constantly changing field that impacts every organization in every industry in the world. You know your actions can have a positive impact around the world.*[8]

FIND OUT MORE

Cybersecurity and Infrastructure Security Agency

www.cisa.gov

The Cybersecurity and Infrastructure Security Agency (CISA) is a US government agency responsible for cybersecurity across the country. The careers link on the CISA website has information on opportunities for high school and college students.

International Information System Security Certification Consortium

www.isc2.org

The International Information System Security Certification Consortium (ISC2) offers training and certification in information system security. ISC2 also offers continuing education for those already working in the field.

OTHER JOBS IN THE TECHNOLOGY INDUSTRY

Penetration Tester

Penetration testers replicate hacker attacks on an organization's systems. Organizations hire penetration testers to find network weaknesses before illegal hackers do. Penetration testers gather information about the system. They find open areas where they can get into the system. Testers go into the system and access data. They report any problems they find to the organization. Once the test is done, the organization makes sure all the accessed data is returned to its original form.

Smart Home Installer

Smart homes have devices connected to the internet. They help automate basic household tasks. Technology controls the lights and temperature. Homes have security devices. These include cameras or automatic locks. In a smart home, music or the oven may be turned on before the owner arrives. These homes often have a voice assistant. Smart home installers are responsible for setting up all the devices in a home. They make sure the devices work together. They also keep everything running smoothly.

Robotics Technician

Robotics technicians install and program industrial robots. These are machines that make work easier and more efficient for different industries. Manufacturers use robots for jobs such as welding or painting. Farmers use robots for picking vegetables. Warehouses use robots to stock shelves. Robotics techs test and maintain the equipment. If there are problems, techs troubleshoot and repair the robot. Some robotics technicians help design and build robots. They also train others to use the robots in the workplace.

Video Game Designer

Video game designers develop ideas for new video games. They use their creative skills to create the game's characters. They tell stories about the characters. And they decide what happens in the game. Designers may build and test game samples. Video game designers often work on a team. A programmer writes the code that makes the game work. The team may also include artists and writers.

GLOSSARY

automatic

working by itself with little or no human help

breach

a break or gap in a security system

hackers

people who try to gain illegal access to computer systems

networks

systems of computers and technical devices connected together

portfolio

a collection of samples of a person's best work

specifications

detailed descriptions of designs and materials needed to make something

troubleshoot

to identify problems in an electronic device

viruses

programs that can damage or disrupt normal computer functions

SOURCE NOTES

CHAPTER ONE: CNC TOOL PROGRAMMER

1. Quoted in "What a CNC Programmer Does & How to Become One," *Universal Technical Institute*, August 4, 2025. www.uti.edu.

2. Quoted in "What a CNC Programmer Does & How to Become One."

CHAPTER TWO: COMPUTER SUPPORT SPECIALIST

3. Quoted in Pius, "Computer Support Specialist—Ben's Answer," *CareerVillage.org*, October 28, 2021. www.careervillage.org.

4. Quoted in "What's It Like to Work in IT Support? Inside Look at the Role," *Herzing College*, June 4, 2021. https://blog.herzing.ca.

CHAPTER THREE: WEBSITE DEVELOPER

5. Quoted in Emily Stevens, "What's a Typical Day in the Life of a Web Developer?" *CareerFoundry*, September 21, 2023. https://careerfoundry.com.

6. Brandon Arnold, "A Day in the Life of a Web Designer/Developer," *Cyphers Agency*, May 25, 2016. https://thecyphersagency.com.

CHAPTER FOUR: CYBERSECURITY ANALYST

7. Quoted in "Ask the Experts: What's Most Rewarding About Your Career in Cyber Security?" *Black Duck*, February 2, 2020. www.blackduck.com.

8. Quoted in "Ask the Experts: What's Most Rewarding About Your Career in Cyber Security?"

INDEX

IMAGE CREDITS

Cover: © Standret/Shutterstock Images
4: Red Line Editorial
5: Red Line Editorial
7: © Daisy Daisy/Shutterstock Images
8: © AlesiaKan/Shutterstock Images
10: © Frame Stock Footage/Shutterstock Images
13: © Narongrit Sritana/iStockphoto
15: © Pixel B/Shutterstock Images
16: © Monkey Business Images/Shutterstock Images
20: © Gorodenkoff/Shutterstock Images
25: © DC Studio/Shutterstock Images
27: © Gorodenkoff/Shutterstock Images
28: © Monkey Business Images/Shutterstock Images
31: © Gorodenkoff/Shutterstock Images
32: © dotshock/Shutterstock Images
37: © AnnaStills/iStockphoto
39: © Insta_Photos/Shutterstock Images
40: © Premreuthai/Shutterstock Images
43: © Orian Lev Ari/Shutterstock Images
44: © Media_Photos/Shutterstock Images
49: © Gorodenkoff/Shutterstock Images
52: © Media_Photos/Shutterstock Images
58 (top): © Roman Samborskyi/Shutterstock Images
58 (bottom): © Suwan-Studio/Shutterstock Images
59 (top): © Chatchawal Phumkaew/Shutterstock Images
59 (bottom): © Frame Stock Footage/Shutterstock Images

ABOUT THE AUTHOR

Cynthia Kennedy Henzel has degrees in education and geography. She has written more than a hundred books for young people. These include fiction and nonfiction on subjects such as geography, science, culture, and history. Henzel enjoys discovering new things—especially discovering how something works or a new solution to a problem. She finds technology interesting to learn about.